Walter Böhm & Peter Siebert

PUBLICATIONS COMPANY

603-609 Castle Peak Road
Kong Nam Industrial Building
10/F, B1, Tsuen Wan
New Territories, Hong Kong

We welcome authors who can help expand our range of books. If you would like to submit material, please feel free to contact us.

We are always on the look-out for new, unpublished photos for this series.
If you have photos or slides or information you feel may be useful to future volumes, please send them to us for possible future publication.
Full photo credits will be given upon publication.

ISBN 962-361-038-6
Printed in Hong Kong

Front Cover

"Target enemy". A two-man German paratrooper MILAN anti-tank team with the VW ILTIS in combat with WIESEL. (Peter Siebert)

Back Cover

The FH-70 from Feldartilleriebtl.555 in a firing position. The FH-70 is the result of a tri-national project among Great Britain, Germany and Italy, for a new 155mm howitzer. At end of 1991, Great Britain has 71, Germany 216 and Italy 164 FH-70s in their inventory. (Peter Siebert)

Introduction

The FRANCO-GERMAN BRIGADE (D/F-Brig.) is really the first European military unit organized for a defensive role that consists of soldiers from different nations.

After the end of Cold War, the free nations tried to find new and better ways to carry out their defense policy.

Today , due to political developments, troops and financial resources are reduced. New organization has become necessary to give the reduced forces a better capability for quick defensive operation.

So the European nations and their allied partners are forced to work together. As a result, the nations have built up bi- or multi-national combat units like the MNAD or D/F-Brig.

While the MNAD is still passing through the trial phase, the D/F-Brig. showed combat readiness during the first brigade scale exercise "ALB 91" in the spring of 1991.

Although we can't compare the D/F-Brig. with the newly founded FRANCO-GERMAN ARMY CORPS, the exemplary performance of D/F Brig. at ALB 91 and the experience it gained show it will help with the difficult task of building-up a bi national unit.

Here are some historical facts about this unusual military unit:

1987 - French President Mitterand and German Chancellor Kohl decide to establish a FRANCO-GERMAN BRIGADE.

1988 - First staff units start their work in order to build up the new brigade structures and organization.

1990 - The brigade build-up phase has finished.

June 91 - During the logistic exercise "BLAUE DONAU" and the brigade scale FTX "ALB 91", the D/F-Brig. appears as a well instructed and highly motivated military unit.

Oct. 91 - Brigade command changes from French General Sengeisen to German General Neubauer.

The D/F-Brig. troops are stationed in southwest Germany. They consist of these nine main units:

<u>Mixed units:</u>

The STAB und STABSKOMPANIE (headquarters company) is a mixed Franco-German unit. The command changes every year between French and German generals.

The VERSORGUNGSBATAILLON (support battalion), the unit responsible for the brigade logistic tasks.

<u>French units:</u>

The AUFKLÄRUNGSKOMPANIE (reconnaissance company) is the first French unit that has been equipped with the brand new Panhard VBL with FN gun or MILAN anti-tank system.

The 3rd HUSSARENREGIMENT (light tank battalion) provides main firepower against an armored enemy. The AMX 10 RC tanks, with superior speed, good cross-country capability, and their 105mm main gun, are a flexible tool against all kinds of enemy attacks. The 3rd HUSSARS are also equipped with the VBL for transportation and observation roles.

The 110th INFANTERIREGIMENT, well equipped with the VAB APC, towed mortars, 20mm field guns and many MILAN anti-tank systems.

<u>German units:</u>

The JÄGERBATAILLON 552 (infantry battalion) is the German version to the 110th Regiment. The main weapon system is the well known FUCHS APC. The battalion is also equipped with many MILAN systems. The Germans mount the MILAN on ILTIS light vehicles to get better mobility. In the future, the ILTIS will be replaced by the Mercedes-built WOLF vehicle. Another unusual combination, the UNIMOG weapon carrier with the FK 20mm field gun mounted on the loading area, allows the troops to achieve quick fire support. The regular trailer is used only for transport over long distances.

The FELDARTILLERIEBATAILLON 555 (field artillery battalion), equipped with FH-70 155mm howitzers and the MAN 7t gl truck, which is used as a tractor and ammotruck.

The PANZERJÄGERKOMPANIE 553 (heavy anti-tank company). Their JAGUAR 1 tanks with the HOT anti-tank rocket system are able to destroy all kinds of armor up to a range of 4,000 meters.

The PANZERPIONIERKOMPANIE (engineers) whose task is to support the brigade's combat units with the BIBER bridge-layer, the DACHS CEV and the SKORPION minelayer.

In addition, all these units have their own small support platoons for direct logistic tasks. So, the 4,200 soldier-strong FRANCO-GERMAN BRIGADE has good firepower and is able to react very flexibly under all conditions.

Acknowledgements:

The authors would like to offer special thanks to Hauptmann Reiberling for his generous support.

Franco-German cooperation during the Feldübung 92. (Peter Siebert)

German infantry with full equipment after a battle. The soldiers are members of Jägerbtl.552 and wear the new uniforms. (Walter Böhm)

French tank commanders from the 3rd HUSSARS planning an attack. The regiment is based in Pforzheim, and is one of the backbones of the D/F-Brig. (Walter Böhm)

German FUCHS commander with the 7.62mm machine gun. This weapon is an improved version of the well-known German model MG 42 of WWII. (Walter Böhm)

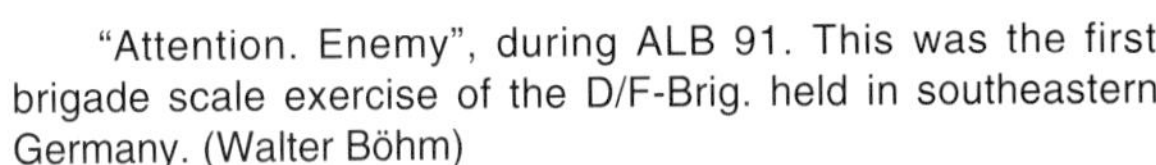

"Attention. Enemy", during ALB 91. This was the first brigade scale exercise of the D/F-Brig. held in southeastern Germany. (Walter Böhm)

PEUGEOT P4 with a machine gun mounted on the hood. For the 3rd HUSSARS, the P4 will be replaced by the PANHARD VBL. (Walter Böhm)

MG troop from Jägerbtl.552 with the new uniform. After a long trial phase, the Bundeswehr will replace the old olive drab uniform. (Peter Siebert)

The German government has reduced the military service time from fifteen to twelve months, and cut down on the exercise activities. So soldiers have only one or two chances to feel what "living in the field" is like. (Peter Siebert)

German infantryman with G3 rifle. The G3 uses the NATO caliber 7.62mm ammunition and has a 20-shot magazine. (Walter Böhm)

The German Hauptmann from Feldartbtl.555 wears the new uniform. A Hauptmann leads a battery with nine FH-70 howitzers. Notice the radio, and the D/F-Brig. patch on the sleeve. (Peter Siebert)

One of the important points during a free land exercise is map reading. These tankers from 3rd HUSSARS have problems finding the right way in the unknown German area. (Peter Siebert)

"Hopeless situation". Notice the special version of the G3. The paratroop version depicted here is only one member of the G3 rifle family; HECKLER & KOCH built many thousands for armies all over the world. (Walter Böhm)

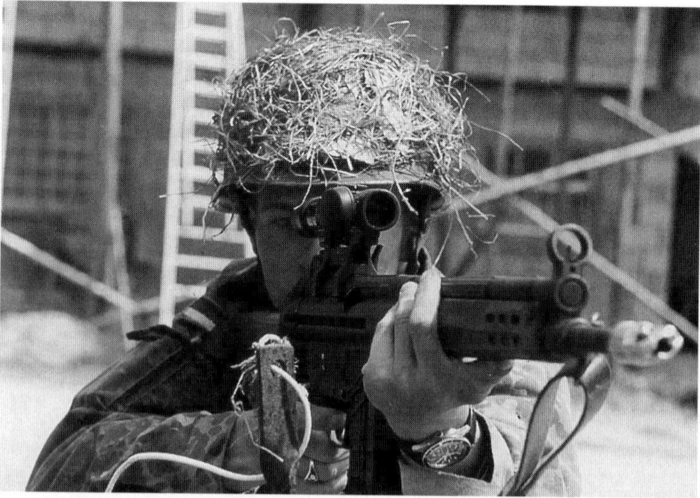

A German sniper with G3 rifle and new-style uniform. This German rifleman is taking part in French sniper training. German unit leaders used every chance to train their soldiers together with their French partners. (Walter Böhm)

A gunner from Feldartbtl. 555 on the seat of this FH-70 howitzer. The German version of FH-70 has a digital display unit(DDU) for the firing data. (Peter Siebert)

Close-up of a MILAN team from the 110th Inf.Reg. The MILAN system project is a British, French and German joint venture. (Walter Böhm)

Another German sniper carries a G3 with a telescopic sight and the new-style weapon belt. The camouflage is similar to some German WWII uniforms. (Walter Böhm)

Infantrymen from Jägerbtl.552. In the Bundeswehr the duty time has been reduced to twelve months since 1990. In this short time, the recruits must learn all their military duties. (Walter Böhm)

DACHS CEV commander, with the new Russian-style tanker helmet. Following a long trial phase, the new tanker Lederhelm has been in service since 1984. This new helmet gives the tank crew better protection in the cramped interior. (Walter Böhm)

The cramped interior of a FUCHS APC gives the fully equipped riflemen little space to move. The FUCHS is a light and highly mobile APC and the workhorse of Jägerbtl.552. (Walter Böhm)

Weapon drill time for French soldiers from the reconnaissance company. New weapons like the FAMAS rifle make things easier than in the past, though. There are a small number of different parts and only some parts must be stripped down, cleaned and reassembled. (Peter Siebert)

Radio operator in the FUCHS. When the company commander leaves the APC, the radio operator maintains contact with the other companies and the battalion commander. (Peter Siebert)

Close-up of a FUCHS commander. The infantrymen have left the FUCHS and the commander protects himself with the machine gun. (Peter Siebert)

German MILAN team from Jägerbtl.552 with its heavy weapon. The MILAN anti-tank missile can destroy all armor up to 3,500m. (Peter Siebert)

French AMX 10RC commander. The 3rd HUSSARS have a long tradition and their history can be traced back to the time of the light cavalry. (Peter Siebert)

A French Captain and a German Hauptmann during COLIBRI 92. All officers in D/F-Brig. must speak two languages; this is the basis for a bi-national unit. (Peter Siebert)

French AMX 10RC gunner from the 3rd HUSSARS. This six-wheeled light tank has a four-man crew. (Peter Siebert)

A German Kradmelder with a special version of the MP (machine pistol). The nickname of this MP is "UZI" (like the Israeli-built MP). This small automatic weapon is reliable and it is the German tank crew's standard weapon. (Peter Siebert)

The FUCHS ambulance is unarmed and can transport four stretches or eight seated wounded, or a combination. This is a very rare shot because there are only four FUCHS ambulance vehicles in the German army's inventory. All are used by Jägerbtl.552, with one per company. (Peter Siebert)

The game is over for this FUCHS. Although these vehicles have an excellent cross-country capability, in this case there was no way out. (Peter Siebert)

A lot of work waits for the crew. After all attempts to rescue the vehicle with another FUCHS provided no results, the company commander called for a DACHS CEV. Four hours later, the FUCHS was towed out of the mud hole without suffering any mechanical damage. (Peter Siebert)

A group of umpires with a P4 observe the activities of a FUCHS crew from Jägerbtl.552. ALB 91 was the D/F-Brigade's first full-scale exercise and so umpires escorted the troops during the day and night. (Peter Siebert)

During COLIBRI 92, parts of the D/F-Brig. play the enemy role. This FUCHS APC takes part in a mechanized forces attack against French and German paratroop positions. (Peter Siebert)

The D/F-Brig. is a mixed unit with a common defense policy. As an external sign of unity, all French and German combat vehicles are painted in the NATO 3-tone camouflage scheme. (Walter Böhm)

FUCHS near a German guest house, following a battle against paratroops during ALB 91. With three rifle companies and a support company, the Jägerbataillon 552 is the largest German unit in D/F-Brig. (Walter Böhm)

Teamwork between an AMX 10RC from 3rd HUSSARS and a FUCHS APC from Jägerbtl.552 during COLIBRI 92. Infantry needs the protection of tanks when operating in this open terrain. (Peter Siebert)

A FUCHS APC in a camouflaged position. The MILAN team left the vehicle to dig in to get a better position. (Peter Siebert)

This FUCHS is marked with red signs for an enemy role. If the FUCHS is used as a cargo carrier, loads of up to four tons can be transported. (Peter Siebert)

Command and communication version of the TPZ FUCHS. This model has an additional 5 KW generator in the rear door. There are radios and mapboards installed inside. When used as a static command post, a high antenna can be fitted at the rear of the hull. (Peter Siebert)

The FUCHS is used as a section vehicle. Ten fully equipped infantrymen can be transported in the rear cargo area, but there is no possibility of fighting within the armor protection. (Peter Siebert)

An infantry company takes a break under protection of a tree line. It is one of the three FUCHS mounted rifle companies. They are backed up by the heavily armed 5th Co. with 120mm mortars and FK 20mm guns. (Walter Böhm)

A FUCHS passes a war memorial in a German town. All personnel of Jägerbtl.552 are volunteers, highly motivated to support the new common defense concept, represented by the D/F-Brigade. (Walter Böhm)

"BIG TROUBLE". This FUCHS followed an "enemy" airborne armored vehicle, a WIESEL during ALB 91. The WIESEL, with a weight of 2.8 tons has no problems; but the FUCHS, with a combat weight of seventeen tons, dragged in the mud totally. (Walter Böhm)

A FUCHS during COLIBRI 92. The vehicle is fully amphibious, being propelled in the water by two propellers. There are no preparations necessary, other than to erect a trim vane. (Peter Siebert)

AMX 10RCs during ALB 91. This first line AMX 10RC from 3rd HUSSARS stopped a counter-attack from an enemy MARDER MICV from the German 10th Tankdivision. (Walter Böhm)

AMX 10RC from 3rd HUSSARS protects a German FUCHS from Jägerbtl.552 during COLIBRI 92. The wheeled vehicles have a great advantage since they can operate silently. (Peter Siebert)

Tankers from 3rd HUSSARS maintain the AMX engine during a break in exercise ALB 91. The crews try to keep their vehicles in good condition in order to be ready anytime. (Walter Böhm)

One special feature of the AMX 10RC light tank is the hydropneumatic suspension. The ground clearance can be changed between 0.21m and 0.6m. The photo shows the vehicle in the lowest position. (Peter Siebert)

"A lot of work": maintenance on an AMX 10RC at the Baumholder training area. In the last production batches, the RENAULT engine was replaced by the better BAUDOIN engine. This gives the tank a larger operating range. (Peter Siebert)

3rd HUSSARS prepare their tanks for a regiment scale exercise. This unit doesn't use the usual camouflage nets; they prefer painted cloths. (Peter Siebert)

AMX 10RC from the 3rd HUSSARS. The regiment has three squadrons each with twelve tanks, supported by a command and a logistic squadron. (Walter Böhm)

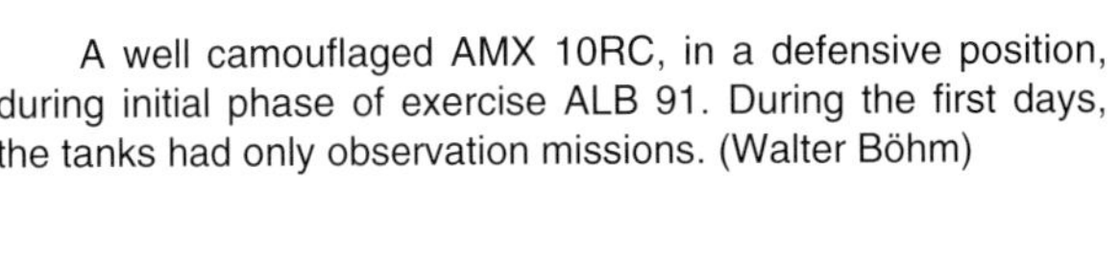

A well camouflaged AMX 10RC, in a defensive position, during initial phase of exercise ALB 91. During the first days, the tanks had only observation missions. (Walter Böhm)

AMX 10RC platoon waits for umpires orders. The platoon was stopped by a fictitious air raid, and the umpires must decide which tanks are destroyed. (Walter Böhm)

AMX 10RC, with the 105mm main gun and a top speed of 100km/h, showed during Gulf War that this light tank is the ideal weapon system for rapid action. (Walter Böhm)

The 3rd HUSSARS take a short break during ALB 91. Umpires decided that the AMX 10RC platoon drove into a minefield. Minefields and destroyed vehicles exist only on the umpires maps. After a waiting period, the "minefield" was cleared and the tanks continued ahead. (Peter Siebert)

"Eye to eye" with a 3rd HUSSARS tank. The 105mm main gun can fire four types of ammunition: practice, HEAT, HE, and APFSDS. The APFSDS can penetrate the NATO heavy tank targets at a range of 2,000m. Thirty-eight rounds of 105mm ammunition are carried, twelve in the turret and twenty-six in the hull. (Peter Siebert)

German farmland has a lot of obstacles. This AMX 10RC skidded into a trench near a manure heap and must be pulled out by another AMX10RC. (Peter Siebert)

The hydropneumatic suspension needs a lot of time for careful maintenance. But on a positive note, crews can use all kinds of trees and shrub lines as natural cover when they lower the whole vehicle. (Peter Siebert)

Some AMX 10RC's during a 3rd HUSSARS regiment exercise. The tactical training starts with exercises at platoon and company levels, and is completed by a full scale regiment or brigade exercise. (Peter Siebert)

An AMX 10RC in a camouflaged observation position. The COTAC fire control system and the laser range finder give the tank good first hit capability. (Peter Siebert)

The MEPHISTO is the anti-tank version of the well-known VAB. The EUROMISSILE-built launcher system on the upper hull can fire four HOT missiles. For reloading, the launcher is retracted into the hull, giving the crew full protection. (Peter Siebert)

The VAB is the standard weapon system in the 110th Inf. Reg. Since 1976, the French Army has received 4,430 VAB vehicles in different versions. The 110th Inf. Reg. used the VABs mostly as section vehicles. (Peter Siebert)

The Infanterie Regiment has 80 VAB APCs in different versions. The commander's gun mount can be fitted with the American Cal.50 MG or the FN 7.62mm machine gun. (Walter Böhm)

The backbone of 110th Inf. Reg. is their twenty-four MILAN systems. Troops can bring them into action by manual transport or mounted on the VAB. (Walter Böhm)

This VAB from RED forces is marked with orange identification panel during exercise COLIBRI 92. COLIBRI is the annual exercise for French and German paratroopers. (Peter Siebert)

The VAB Sanitaire is the unarmed ambulance version of the VAB family. It carries first aid equipment and a mix of stretchers, and seats for the wounded. In addition to the red cross signs, we can identify the Sanitaire from the additional air conditioning system over the troop compartment. (Peter Siebert)

Two VAB with MILAN from the 110th Inf.Reg. in an ambush position. The MILAN is a semi-automatic wire-controlled anti-tank missile. (Walter Böhm)

This VAB is used as a command post for the 110th Inf. Reg. commanders. With the additional radio equipment, this VAB is a communications center, too. (Walter Böhm)

French soldiers prepare for a raid in a German town. Urban warfare was one of the main points during exercise ALB 91. The troops have a lot of problems bringing their different tactics together. (Peter Siebert)

During a battle there is no time to camouflage the VAB with nets or shrubs, so the crew uses the shadow of a German farmhouse roof for protection against enemy jets. (Peter Siebert)

Reflections from the windscreen are perfidious signs for the enemy. On this VAB, the windscreen protection shields are closed. (Peter Siebert)

Close-up of a VAB mounted MILAN. On a trial basis, the French Army fitted some VABs with the EUROMISSILE MILAN compact turret. But the testing had no satisfactory results and the project was canceled. (Walter Böhm)

A VAB from a maintenance section with a big box for additional tools and spare parts. (Walter Böhm)

VABs from 110th Inf.Reg., attractively painted in the NATO 3-color camouflage. These VAB s are fitted with the CB 127 gun ring shield for the M2 Cal.50 MG. (Peter Siebert)

German soldier from the HQ.Co. with the French FAMAS rifle. This unit was the first to be issued with this new weapon. It is a nice rifle but the 5.56mm ammo cannot be used in the German 7.62 caliber G3 rifle. (Walter Böhm)

French crews on weapon drill. Note the metal badges with the French and German national colors on the beret. (Walter Böhm)

The German Jägerbataillon 552 packs an impressive punch. It is equipped with forty FUCHS APCs, twelve MILAN missile launchers, six 120mm mortars and six FK 20mm guns. (Walter Böhm)

FUCHS APCs and the crews are ready for action. Ten minutes later, the enemy (played by German paratroopers) surprised this company with a raid. (Walter Böhm)

It is not easy to handle a big armored vehicle like the FUCHS in an urban area. Here the infantry company gets the order to search and destroy the enemy paratroopers which landed some minutes ago. (Walter Böhm)

FUCHS with infantry during ALB 91. The troops have left their "battle taxi". (Walter Böhm)

This AMX 10RC has the suspension in position for maximum ground clearance of 0.6m, normally used only for amphibious operations. (Peter Siebert)

AMX 10RC in German woodland near Merzig, during the first days of COLIBRI 92. COLIBRI was a mixed exercise, in open land and training areas. (Peter Siebert)

Although a wheeled vehicle, the AMX 10RC has tank dimensions and needs well instructed drivers. The long gun barrel makes the tank particularly hard to handle in small towns or woodland areas. (Peter Siebert)

In the French Army, the AMX 10RC is normally used as a reconnaissance vehicle. The lack of a real tank battalion in the D/F-Brig. makes it necessary that the 3rd HUSSARS operate as heavy forces. (Walter Böhm)

The low silhouette gives the AMX 10RC good passive protection. For a quick escape there are two smoke dischargers mounted on either side of the turret. (Peter Siebert)

Feldübung 92 was held near Baumholder training area. There the 3rd HUSSARS had to operate in an inaccessible low mountain range area. (Walter Böhm)

"Infantry go ahead": soldiers from the 110th Inf.Reg. leave their VAB during ALB 91. The VAB has a crew of two (driver and commander) and can transport up to ten fully-equipped soldiers. The soldiers enter and leave the VAB by a double door on the vehicle's rear. (Peter Siebert)

A tank platoon from 3rd HUSSARS during the preparation phase of exercise COLIBRI 92. The 105mm main gun, a range of 1000km, and a speed up to 100km/h are advantages of the AMX 10RC proven during the Gulf War. (Peter Siebert)

The VAB is also used to transport MILAN anti-tank teams. These vehicles have no special modifications, only some additional racks in the inside to hold MILAN rockets. There is a small mount between the rear hatches for the MILAN launcher. (Peter Siebert)

The VAB (Vehicule de l'Avant Blinde) is a fully amphibious wheeled APC. Before water operations, a folded trim vane on the front must be erected, and the two water jets must be activated by the driver. (Peter Siebert)

Some VAB MEPHISTOs became part of the 3rd HUSSARS for Feldübung 92, but this was only a test to give the commanders experience with tank destroyers. (Walter Böhm)

Another unusual vehicle in 5/552's inventory is the UNIMOG U 1300L with the FK 20mm mounted on the loading platform. The RHEINMETALL-built FK 20mm can be used for air defense and ground support. (Peter Siebert)

The German anti-tank teams from 5th Co., Jägerbtl.552 like their ILTIS mounted MILAN. This suitable combination gives the crew high mobility and excellent cross country capability, the best prerequisites for shoot and hide tactics. (Peter Siebert)

The new PANHARD VBL with MILAN system. The MILAN launcher can be fitted on the front or rear hatches. There are six reserve missiles in the vehicle's interior. (Peter Siebert)

Each reconnaissance platoon has three sub-units of two VBLs each, a MILAN missile unit with two MILAN VBLs and a command group. (Walter Böhm)

The German FK 20mm gun on the regular trailer. This gun, built by RHEINMETALL, is also used in the new WIESEL airborne armored vehicle. (Walter Böhm)

The JAGUAR tank destroyers are rebuilt JAGDPANZER RAKETEs. The main modifications are the HOT launchers and the add-on armor on the front and sides of the hull. (Peter Siebert)

Many preparations are necessary before a live firing exercise can start. A new family of a ammunition has been developed especially for the FH-70 with a range of 24,700m for the standard projectile. After preparation, the FH-70 can fire at a rate of up to ten rounds/minute. (Peter Siebert)

A VAB with an additional maintenance rack. Vacant of seats, the cargo area is free to transport a lot of tools and welding equipment. In the cargo area, loads of up to two tons can be transported. (Walter Böhm)

The BERLIET GBC 8KT 4-ton truck is the French Army's workhorse. Since the late 1950s, more than 18,000 trucks have been produced. Notice the gun mount for the Cal.50 MG on the left truck's cab. (Peter Siebert)

The BERLIET GBC 8KT is also used as a light wrecker. Besides the cargo and light wrecker version, there is a fuel tanker and a compressor carrier based on the GBC 8KT. (Walter Böhm)

The BERLIET GBC 8KT light wrecker. With an overhead gantry, it is used to lift light vehicles and change engines, gear boxes or other components. (Peter Siebert)

A SIMCA UNIC F594WML 3-ton truck from the 3rd HUSSARS maintenance section. This old vehicle will be replaced by the RENAULT TRM 2000 in the near future. (Peter Siebert)

In the French Army inventory are some different variants of the SIMCA UNIC: a cargo truck, van-type body, tanker, dumptruck, compressor carrier and a wrecker. (Peter Siebert)

A long column passes through a small German town during the initial phase of ALB 91. Shown here are BERLIET TBU 1GLD wreckers from the D/F-Brig. support battalion. (Walter Böhm)

In the near future all HOTCHKISS-built M201 "Jeeps" will be replaced by the PEUGEOT P4. Over 40,000 M201s were built under licence from WILLYS between 1953 and 1969. (Peter Siebert)

The PEUGEOT P4 is the replacement for the M 201 "Jeep". Since 1987 the French Army has ordered 15,000 vehicles. The "Umpire" P4 shown here is the standard model, but there are other versions with different gun mounts and the MILAN weapon carrier. (Peter Siebert)

The lack of an effective maneuver simulation system for vehicles and troops makes a large umpire force necessary. This umpire team uses the PEUGEOT P4 and the RENAULT TRM 2000 truck. (Peter Siebert)

The unique variation of the VW-built ILTIS is the ambulance model. Normally, two stretchers can be carried in addition to a crew of two. (Peter Siebert)

This is the end of service for the VW-built ILTIS light vehicles in Feldartbtl.555. The ILTIS will be replaced by the MERCEDES 250G. (Walter Böhm)

The MILAN is the standard anti-tank weapon in the D/F-Brig. Here it is mounted on an ILTIS from the 5/552 Jägerbtl. to get better mobility. (Walter Böhm)

"Panzerknackers" from 5th Co., Jägerbtl.552 in an ambush position. The MILAN anti-tank team consists of a crew of two soldiers, a driver and a MILAN gunner. The teams work separately and get their orders directly from the company commander. (Walter Böhm)

The German Army replaced the VW-built ILTIS by the military version of MERCEDES 250G. This vehicle now has the official name "WOLF". The WOLF shown here is a part of enemy "RED" forces during exercise COLIBRI 92. (Peter Siebert)

A MAN 7tgl truck from Panzerpionierkompanie 550 tows a MIVS trailer. With the MIVS (Minen Verlege System-mine laying system), anti-tank mines can be placed on the surface (800 mines/hour) or under ground with a plough (500 mines/hour). (Peter Siebert)

The MAN 10tgl trucks used for general logistic support in Feldartilleriebtl.555. Loading operations are easy with the cargo area-mounted 1-ton crane. (Peter Siebert)

Notice the different painting schemes of the MAN field kitchen trucks. Maintenance of these old vehicles becomes more and more expensive, so they will be replaced by towed field kitchen trailers. (Peter Siebert)

ELEFANT tank transporter from the support battalion. The SLT 50-2 can carry all kinds of vehicles, including the LEOPARD 2 tank. A dual winch with a capacity of 17 tons is mounted on the tractor. (Peter Siebert)

A 30-year-old workhorse, the MAGIRUS JUPITER 7-ton wrecker. With a crane capacity of four tons, this vehicle is no longer up to date, and will be replaced by the new LIEBHERR wreckers. (Peter Siebert)

After the end of exercise ALB 91, all wheeled vehicles used civil roads to travel home. This VBL was used in a convoy controller role near a dangerous road crossing. (Peter Siebert)

The PANHARD VBL (VEHICULE BLINDE LEGERE) is a light armored amphibious vehicle. The squadron has thirty-six VBLs distributed among three reconnaissance platoons. (Walter Böhm)

VBL: fast-quiet-small, ideal for reconnaissance. In the past, reconnaissance vehicles became bigger and heavier, (see M3 or LUCHS). With the VBL, a new generation of light reconnaissance vehicles enters into service. (Peter Siebert)

ALB 91 was held in a low mountain range, with a lot of valleys. Light vehicles like the VBL can use all the advantages of this area. (Peter Siebert)

Close-up of the VBL's backside. Notice the camouflaged inside of the door. The reconnaissance squadron, with the VBL, is the D/F-Brig.'s eyes and ears. (Walter Böhm)

Most VBLs are armed with the FN 7.62mm MG on a mount over the commander's hatch. In addition to the FAMAS rifles for the crew, three thousand rounds of ammunition are carried. (Peter Siebert)

The first deliveries of VBLs arrived in the French Army in 1989. The reconnaissance company in D/F-Brig. was one of the first units equipped with the new vehicles. (Peter Siebert)

For air defense, the U 1300L can be fitted with a gun mount for the 7.62mm MG above the cab. Notice the air inlet on the cab's right side which allows the UNIMOG to ford rivers up to 1.2m in depth without preparation. (Walter Böhm)

The heavy 5th Co., Jägerbtl.552 has a lot of miscellaneous equipment. For light engineer operations, this company uses a special version of the UNIMOG truck series. It can used as a loader, dozer or excavator. (Peter Siebert)

The UNIMOG with the 20mm FK. When the hood is in the normal position, it is very difficult for the enemy to identify the truck as a weapons carrier. (Peter Siebert)

For transport over far distances, the FK 20mm is mounted on a two-wheel trailer, and towed by the UNIMOG U 1300L or all other vehicles over the one-ton class. (Walter Böhm)

The French FK 20mm is normally towed by a truck or the VAB on a two-wheel trailer. Here it is used in the anti-aircraft role. (Peter Siebert)

French soldiers from the 110th Inf. Reg. prepare the 120mm mortar for a firing mission. The THOMSON BRANDT-built mortar is towed by a VTM-named version of the VAB. The VTM transports the mortar crew and seventy mortar rounds. (Peter Siebert)

The three rifle companies from the 110th Inf.Reg. are backed up by a heavy support company with six 120mm mortars. This mortar crew, with FAMAS rifles, camouflage their mortar during the Feldübung 92. (Peter Siebert)

A well camouflaged Beobachtungspanzer M113A1G. This German conversion of the APC is created to give the forward observers better protection, greater mobility and more room for the radio and fire control equipment. (Peter Siebert)

The German Army decided that the amphibious capacity of the M113 series was no longer necessary. A new-styled stowage basket replaced the trim vane on this M113A1G from the HQ company. (Peter Siebert)

In the front is the RATAC fire control vehicle; in the back is a forward observer. Both vehicles are based on the M113A1G. Each battery has six forward observer M113A1Gs. The 1st Co. Feldartilleriebtl.555 uses the RATAC radar system based on the M113A1G for battlefield surveillance and location. The RATAC antenna is normally covered in the big box on the upper hull. When it's operational, the antenna can be erected by a telescopic mast. (Walter Böhm)

Close-up of the RATAC fire control M113A1G. In the battalion's command battery, we find two RATAC systems. The radar has four operating modes: ground surveillance, acquisition and identification, angular deviation, measurement, and automatic tracking. (Walter Böhm)

In the D/F-Brig. Headquarters Company, we find some M577A1G command post vehicles. Every vehicle carries additional equipment to build a tent at the rear of the vehicle. For the brigade staff , two or three M577A1Gs can be connected by these tents, back to back, or in the form of a "T". (Peter Siebert)

The M577A1G command post vehicle from HQ company is also equipped with the new stowage basket. Other German modifications are changed lights and a new track. (Peter Siebert)

SKORPION minelayer with covered mine racks. The SKORPION is derived from the US M548 cargo carrier, and transports six launching magazines which contain 100 AT-2 anti-tank mines each. (Walter Böhm)

The JAGUAR 1 tank destroyer with the HOT missile system can destroy all armor up to 4,000m. Panzerjägerkompanie 550, based in Stretten, has a total strength of twelve JAGUAR 1s and one M88 recovery tank. (Walter Böhm)

A platoon of JAGUAR 1 from Panzerjägerkompanie 550 in a German town. The photo gives a good impression of the low profile of this tank. (Peter Siebert)

JAGUAR 1 crew take a break during final phase of ALB 91. The add-on armor is bolted on by rubber mountings on the vehicle, and acts as a buffer between the armor and the hull. (Peter Siebert)

The JAGUAR 1 is the standard combat vehicle of Panzerjägerkompanie 550. The tank destroyer uses the HOT missile system has one 7.62mm MG in the hull and another on the commander's cupola for air defense. (Peter Siebert)

A modified KANONENJAGDPANZER now used as a combat zone observer and a command post vehicle. The 90mm gun was removed and an additional periscope installed. (Peter Siebert)

There are 105 BIBER bridgelayers in Bundeswehr inventory. The main advantage of these is that the bridge is extended horizontally rather than vertically like most other bridgelayers. (Peter Siebert)

BIBER bridgelayer from Panzerpionierkompanie 550. With a bridge length of 22m, the BIBER can close gaps up to 21m. More than 60% of all water obstacles in Germany are this width. (Walter Böhm)

During the 1960s, Bundeswehr bought some M88 recovery tanks for the M48A2 tank battalions. After force reduction, some M88s went to other armored units. So, Panzerjägerkompanie 550 and the support battalion are equipped with these "old warriors". (Peter Siebert)

The M88 is the heaviest tank in D/F-Brig.'s inventory. The American M88, built by BMY, is the unique foreign vehicle in the brigade. (Walter Böhm)

An old M88 used in the D/F-Brig.'s support battalion. When using the bulldozer blade as stabilizer, the A-type boom can lift vehicles up to eighteen tons. (Peter Siebert)

This M88 works in the Panzerjägerkompanie maintenance section. The main drawback of the M88 is its age and the missing NBC system. (Walter Böhm)

The DACHS combat engineer vehicle is the newest member of the LEOPARD tank family. The DACHS vehicles are rebuilt Bergepanzer or Pionierpanzer. The main modifications are the new styled dozer blade and the excavator on the telescopic arm. The crew can use the excavator by a remote control unit from outside the vehicle too. (Peter Siebert)

The big excavator shovel is an excellent tool to build up fire positions. In the NATO, only the Bundeswehr and the 4th Canadian Brigade use the DACHS CEV. (Walter Böhm)

This front view of the German DACHS CEV shows the new style bulldozer blade and the extendable hydraulic arm. A new commander's cupola was installed, too. (Peter Siebert)

Used as an excavator, the DACHS has a capacity of 140 cbm/h. In addition to carrying a lot of tools, the DACHS is equipped with a welding and cutting kit. (Peter Siebert)

Close-up of the FH-70 155mm howitzer with the MAN 7tgl truck. Although the howitzer is equipped with the 70hp APU (Auxiliary Power Unit), for all long distance movement, the MAN 7tgl truck is used as a tractor. (Peter Siebert)

This FH-70 was unhooked and driven by itself into a firing position. Before it is ready for action the gun's main wheels must be raised, its barrel rotated to the front, and the trails spread and equipped with the spades. (Peter Siebert)

The FH-70 is self-propelled by the APU 70hp engine. With the APU, the howitzer can drive short distances with a speed of up to 16km/h. In firing position, the APU provides power for all the hydraulic systems like steering, lowering and raising main and steering wheels. But in an emergency all systems can be served by a manually operated hydraulic pump. (Peter Siebert)

Quick movement from one firing position to another needs well trained crew. A good 8-man crew needs only two minutes to uncouple, drive into position and go into action. (Peter Siebert)

Although the FH-70 is a good, reliable weapon system, there is one big flaw - no protection for the crew. Especially under NBC conditions, this is no longer an up to date weapon system. (Peter Siebert)

"STEPS AHEAD.....ALWAYS"

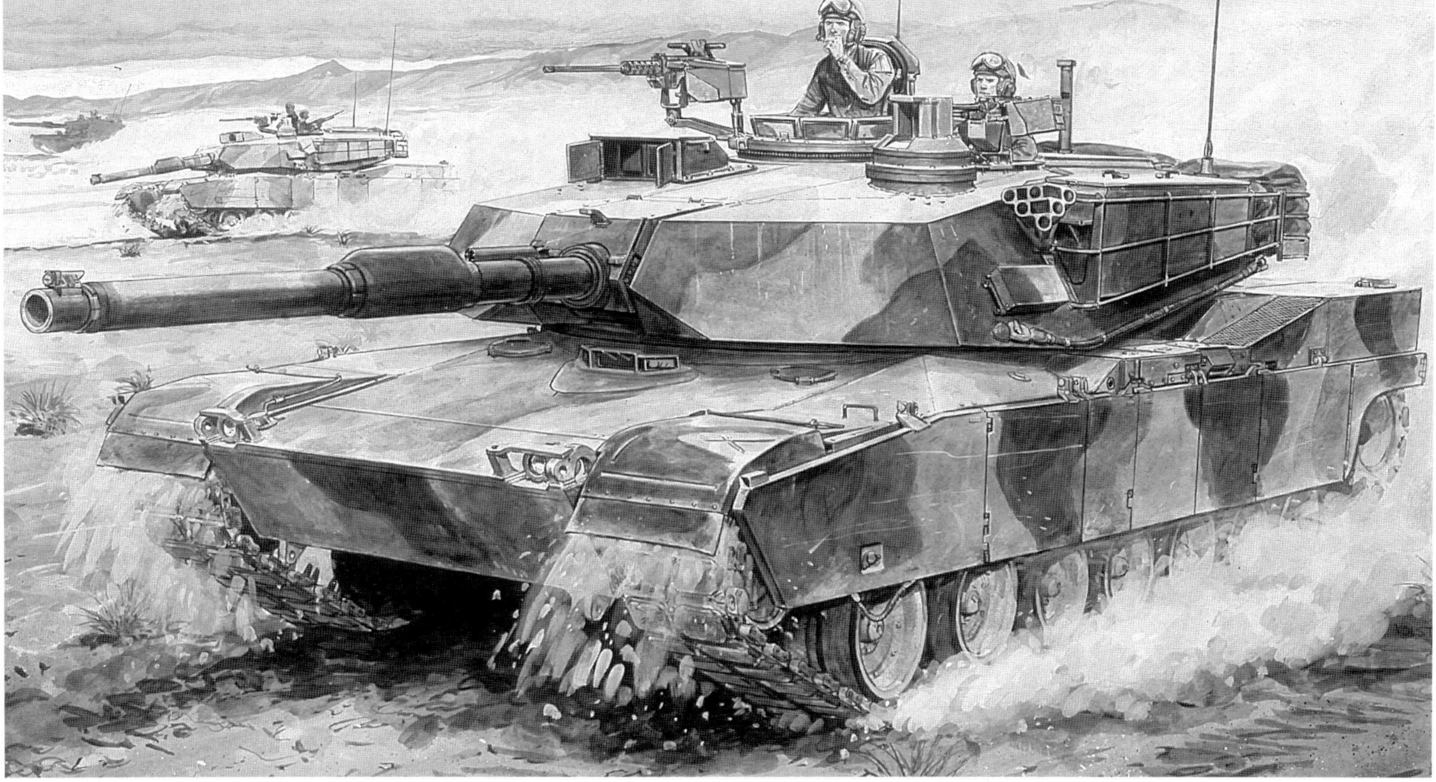

3524 M1A2 ABRAMS

3521 ZSU-23-4V1 SHILKA

3522 NATO MULTIPLE LAUNCH ROCKET SYSTEM

3021 GERMAN PARATROOPERS

3014 FRENCH FOREIGN LEGION

1:35 Scale Series *by*

1007 Leopard 1 and 2: The Spearheads of the West German Armored Forces
Thomas Laber

1018 Modern German Panzergrenadiers: Germany's Mechanized Infantry
Michael Jerchel

2010 The Balkans at War: Yugoslavia Divided 1991
Eric Micheletti

2011 USAREUR: United States Army in Europe
Michael Jerchel

2012 Certain Shield: The Multinational Airmobile Division on Exercise
Michael Jerchel

2013 T-54, T-55 and T-62
Steven J. Zaloga

1031 T-64 and T-80
Steven J. Zaloga

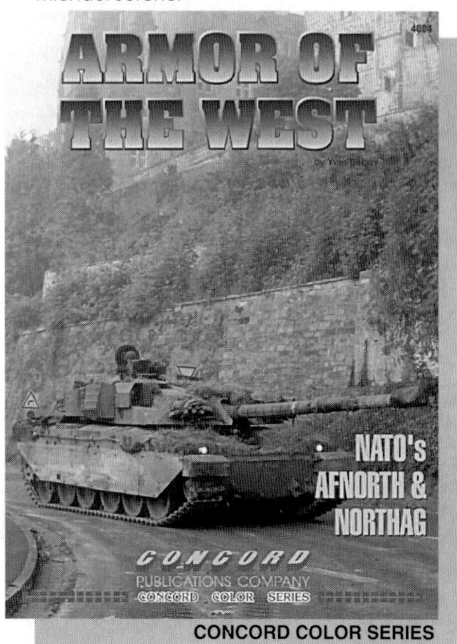

CONCORD COLOR SERIES

4004 Armor of the West: (1) NATO's AFNORTH & NORTHAG
Yves Debay

PUBLICATIONS COMPANY